SPIRAL GALAXY

DISCOVERING HIDDEN FACTS AND MYSTERIES OF THESE GALAXIES

KRISTIE FLEMING

TABLE OF CONTENT

Introduction to Galaxies

Our journey through the cosmos begins with a profound exploration of galaxies, those vast systems of stars, gas, and dust that adorn the universe. In this chapter, we will embark on a captivating journey, delving into the diverse shapes that galaxies exhibit, understanding the significance of spiral galaxies, and gazing upon the grandeur of our home, the Milky Way.

Overview of Different Galaxy Shapes

Galaxies, the cosmic building blocks of the universe, manifest in a variety of captivating shapes. Imagine the universe as a vast art gallery, with each galaxy contributing a unique masterpiece. Among the myriad forms, three primary shapes stand out: spiral, elliptical, and irregular.

Spiral Galaxies: The Twisted Beauties
Spiral galaxies, with their mesmerizing spiral arms and central bulges, dominate the cosmic tapestry. These galaxies constitute approximately 72% of the galaxies that astronomers have observed. Characterized by a central bulge encircled by a rotating

disk, spiral galaxies showcase a breathtaking dance of stars. The Milky Way, our cosmic abode, proudly dons the mantle of a spiral galaxy, a testament to the elegance of this galactic form.

Elliptical Galaxies: The Enigmatic Ellipses

In contrast to the graceful spirals, elliptical galaxies feature a more rounded, elliptical shape. Comprising older, dimmer stars, these galaxies lack the distinctive spiral arms. Their appearance suggests a cosmic serenity, yet they hold mysteries regarding their formation and evolution. Understanding the role of elliptical galaxies enriches our comprehension of the dynamic processes sculpting the universe.

Irregular Galaxies: Cosmic Chaos

Irregular galaxies defy the order observed in spirals and ellipticals. Their shapes lack the symmetrical elegance seen in their counterparts, displaying a chaotic arrangement of stars, gas, and dust. While irregular in form, these galaxies contribute to the cosmic mosaic, adding an element of unpredictability to the grand narrative of the universe.

Importance of Spiral Galaxies

Amidst the cosmic diversity, spiral galaxies stand out as celestial wonders with profound importance. These captivating structures hold the key to understanding the dynamic processes shaping galaxies and, by extension, the universe itself.

Spiral galaxies, with their intricate spiral arms and central bulges, serve as cosmic laboratories. The rotation of stars within the galactic disk and the mysterious formation of spiral arms present astronomers with puzzles waiting to be unraveled. As we explore the significance of spiral galaxies, we gain insights into the mechanisms driving galactic evolution.

One of the defining features of spiral galaxies is their youthfulness. These galaxies host a population of hot, young stars, illuminating the cosmic stage with their brilliance. The interplay between these young stars, the older ones in the central bulge, and the potential supermassive black holes at the galactic core paints a vibrant picture of the complex dynamics within spiral galaxies.

Beyond their celestial beauty, spiral galaxies contribute to our understanding of the broader universe. They represent a crucial piece in the galactic classification puzzle, bridging the gap between the structured spirals and the enigmatic ellipticals. As we delve into the mysteries of these galaxies, we unlock insights into the cosmic ballet that shapes the destiny of galaxies across the cosmos.

The Milky Way as a Spiral Galaxy

As we gaze into the night sky, the Milky Way unfolds above us like a celestial river, weaving its way through the cosmos. Our home galaxy, the Milky Way, serves as both a beacon in the night and a fascinating subject of study for astronomers.

The Milky Way, a sprawling spiral galaxy, encapsulates the essence of our cosmic neighborhood. Nestled within its spiral arms, our solar system finds its place, orbiting the galactic center. The central bulge of the Milky Way, composed of older, dimmer stars, conceals the potential presence of a supermassive black hole, adding an element of cosmic intrigue to our galactic residence.

As we contemplate the Milky Way's spiral structure, we recognize the intricate dance of stars within its rotating disk. The spiral arms, adorned with gas, dust, and young stars, create a celestial panorama that has inspired awe and wonder throughout human history. The Milky Way, with its cosmic splendor, beckons us to explore the mysteries that lie within its vast expanse.

Characteristics of Spiral Galaxies

Our exploration into the cosmic wonders of spiral galaxies deepens as we unravel their distinctive characteristics. In this chapter, we will delve into the mesmerizing features that define these celestial entities, from the central bulge and rotating disk to the intricate spiral arms and the enigmatic presence of bar structures within the cosmic tapestry.

Central Bulge and Rotating Disk

At the heart of every spiral galaxy lies a central bulge, a dense region that holds the secrets of the galaxy's past and present. This bulge, akin to the cosmic core, is a congregation of older, dimmer stars. Within this stellar haven, astronomers suspect the presence of a supermassive black hole, a gravitational behemoth that influences the dynamics of the entire galaxy.

Surrounding the central bulge, like a cosmic dance floor, is the rotating disk—a vast expanse that hosts the majority of a spiral galaxy's stars. This disk, in perpetual motion, plays a crucial role in shaping the galaxy's overall structure. The rotation of stars within the disk, influenced by gravitational forces, contributes to the intricate ballet of a spiral galaxy.

The combination of the central bulge and the rotating disk forms the foundational architecture of spiral galaxies. This dual structure sets the stage for the captivating cosmic drama that unfolds within these galaxies. The central bulge, with its collection of older stars, stands as a testament to the galaxy's evolutionary history, while the dynamic disk embodies the ongoing celestial choreography.

Spiral Arms and Galactic Structure

The defining feature that transforms a standard galactic ensemble into a spiral masterpiece is the emergence of spiral arms. These majestic arms, winding their way through the galactic disk, create a visual spectacle that has captivated astronomers and stargazers alike.

Spiral arms are not static features; they are dynamic and ever-changing. These arms house a wealth of cosmic elements—gas, dust, and a plethora of young, luminous stars. The intricate patterns formed by these arms contribute to the galaxy's overall structure, painting a celestial canvas that tells the story of stellar birth and evolution.

The formation of spiral arms remains one of the great puzzles of galactic astrophysics. One prevailing theory suggests the presence of density waves traveling through the outer disk. These waves, akin to ripples in a galactic pond, could be responsible for the emergence and persistence of the spiral arms. Galactic encounters, where two galaxies interact and influence each other gravitationally, are another potential factor contributing to the formation of these captivating features.

The beauty of spiral arms extends beyond their visual allure. They serve as stellar nurseries, giving birth to young stars that shine brightly before their eventual demise. The life cycle of stars within these arms adds a dynamic dimension to the galactic structure, creating a cosmic symphony that resonates across the vastness of space.

Bar Structures in Spiral Galaxies

While many spiral galaxies showcase the classic spiral arm configuration, a significant number also harbor an additional feature—the presence of bar structures. These bars, composed of stars arranged in a straight bar shape, connect the galactic arms, altering the traditional spiral appearance.

Approximately two-thirds of all observed spiral galaxies exhibit bar structures, including our own Milky Way. Unlike the arms that lead directly into the galactic center, bars provide a unique structure, connecting to the ends of the galactic arms. At the center of these bars lies the galactic nucleus, a region of heightened stellar density.

Barred spiral galaxies add a layer of complexity to the classification of spiral galaxies. The presence or absence of a bar contributes to the variety observed within this galactic class. The Milky Way, for example, is believed to be a barred spiral galaxy, with a central bar structure enhancing its cosmic allure.

Understanding the role of bar structures within spiral galaxies is essential for unraveling the galactic evolution. The gravitational influence of bars can affect the distribution of stars and gas within the galactic disk, influencing the overall dynamics of the spiral system. Observations of spiral galaxies with and without bars contribute to our comprehension of the factors shaping these cosmic wonders.

Formation and Evolution

As we journey through the cosmos, the narrative of spiral galaxies unfolds as a tale of cosmic creation and transformation. In this chapter, we embark on a quest to understand the intricacies of the formation and evolution of spiral galaxies, exploring the enigmatic puzzle of spiral arm formation, the cosmic dance of galactic encounters and density waves, and the mysterious transition into elliptical galaxies.

The Puzzle of Spiral Arm Formation

The mesmerizing spiral arms that define these celestial entities remain a captivating enigma in the realm of galactic astrophysics. How do these graceful arms emerge, persist, and evolve within the dynamic landscape of a spiral galaxy?

One prevailing theory proposes the existence of density waves traversing the galactic disk as the orchestrators of spiral arm formation. Picture these density waves as cosmic ripples, propagating through the outskirts of the galactic disk. As stars orbit the galactic center, they encounter these density waves, leading to the compression of gas and dust, ultimately giving rise to the distinctive spiral patterns we observe.

However, the puzzle deepens when considering the transient nature of spiral arms. Unlike static structures, the arms are dynamic and subject to constant change. The mechanisms responsible for sustaining and rejuvenating these arms over cosmic timescales present a challenge to astronomers.

Galactic encounters, where two galaxies interact gravitationally, represent another intriguing avenue for the formation and disruption of spiral arms. The gravitational tango between galaxies can induce disturbances in their structures, leading to the creation or modification of spiral patterns. The interplay between cosmic dance partners influences not only the appearance of spiral arms but also the overall evolution of the interacting galaxies.

Unraveling the puzzle of spiral arm formation is crucial for a comprehensive understanding of the dynamic processes shaping spiral galaxies. As we explore this cosmic mystery, we venture deeper into the heart of galactic evolution.

Galactic Encounters and Density Waves

The vastness of space is not an empty void; it is a bustling arena where galaxies interact, influencing each other in a gravitational ballet that shapes their destinies. Galactic encounters, where two galaxies come into close proximity, represent pivotal moments in the life of spiral galaxies.

These encounters are not gentle affairs; they are gravitational duels that can leave a lasting impact on the structure and dynamics of the participating galaxies. When galaxies engage in this cosmic dance, their gravitational forces induce ripples, or density waves, within the galactic disks. These waves, akin to celestial tsunamis, create regions of compression and rarefaction, setting the stage for the emergence of spiral arms.

The influence of galactic encounters extends beyond the immediate participants. Even a smaller galaxy can affect the structure of a larger one, leaving an imprint on its spiral arms and overall galactic morphology. These gravitational interactions contribute to the diversity observed among spiral galaxies, with each encounter adding a unique chapter to the galactic narrative.

Density waves, set in motion by encounters or internal mechanisms, serve as cosmic sculptors, shaping the gas, dust, and stars within the galactic disk. The dynamic interplay between density waves and the rotation of stars within the disk creates the intricate patterns we recognize as spiral arms. As we decipher the language of density waves, we gain insights into the mechanisms governing the vibrant tapestry of spiral galaxies.

Evolution into Elliptical Galaxies

While spiral galaxies showcase the splendor of cosmic spirals, their journey through the cosmos is not static. As these galaxies age, they undergo transformations that lead to a diverse array of galactic forms. One intriguing avenue of evolution involves the transition from spiral galaxies to elliptical galaxies.

Elliptical galaxies, characterized by their rounded, elliptical shapes, present a stark contrast to the graceful spirals. The transition from a spiral to an elliptical galaxy is a cosmic metamorphosis that has captured the attention of astronomers, posing questions about the factors driving this transformation.

The key lies in understanding the populations of stars within these galaxies. Spiral galaxies, with their spiral arms and a mix of young and old stars, represent dynamic and evolving systems. As these galaxies age, their stars undergo a natural life cycle, with young stars shining brightly before evolving into older, dimmer stars.

Elliptical galaxies, on the other hand, consist predominantly of older stars. Their elliptical shapes suggest a more relaxed and settled state compared to the dynamic spirals. The transition from a spiral to an elliptical galaxy involves a gradual shift in the stellar population, with the dominance of older stars creating the characteristic appearance of elliptical galaxies.

The processes driving this transition remain a subject of active research. Galactic mergers, where two galaxies combine into a single, larger entity, represent a plausible mechanism. During these mergers, the gravitational interactions between galaxies can disrupt the ordered structure of spiral arms, leading to the formation of a more spheroidal, elliptical galaxy.

As we explore the evolution into elliptical galaxies, we gain insights into the cosmic life cycle of spiral galaxies. The intricate dance of gravitational forces, stellar evolution, and galactic interactions paints a dynamic portrait of the evolving universe. The transition from spirals to ellipticals adds a layer of complexity to the galactic narrative, revealing the diverse pathways galaxies traverse through the cosmic ages.

Notable Spiral Galaxies

In the vast cosmic tapestry, certain spiral galaxies stand out as celestial marvels, captivating astronomers and stargazers alike. In this chapter, we embark on a journey to explore the remarkable features and cosmic significance of notable spiral galaxies, including the colossal NGC 6872, the ancient A1689B11, and the familiar Milky Way with its distinctive bar structure.

NGC 6872: A Giant Spiral

NGC 6872, residing in the southern constellation Pavo, commands attention as one of the largest spiral galaxies known to astronomers. Stretching its arms across an astounding 522,000 light-years, NGC 6872 dwarfs our own Milky Way, emphasizing the cosmic scale of these stellar systems.

The grandeur of NGC 6872 extends beyond its size. This giant spiral galaxy showcases a striking example of spiral arms unfurling gracefully from a central bulge. As we delve into the details of NGC 6872, we encounter a cosmic masterpiece, a testament to the intricate dance of stars within its colossal disk.

Observations of NGC 6872 reveal the complexity of its galactic structure. The spiral arms, adorned with stars, gas, and dust, create a mesmerizing display of cosmic artistry. The dynamics within NGC 6872, including the rotation of stars in its disk and the potential presence of a central supermassive black hole, provide valuable insights into the behavior of giant spiral galaxies.

As astronomers study NGC 6872, they unravel the mysteries of these colossal cosmic entities. The interactions between stars, the influence of gravitational forces, and the evolution of such massive spirals contribute to our broader understanding of galactic dynamics. NGC 6872, with its grandeur and complexity, stands as a beacon in the exploration of the universe's most magnificent structures.

A1689B11: An Ancient Spiral Galaxy

In the cosmic theater, where time is measured in billions of years, the discovery of an ancient spiral galaxy adds a layer of intrigue to our understanding of galactic evolution. A1689B11, located in the constellation Sextans, invites us to peer into the depths of cosmic history.

A1689B11 earned its distinction as an ancient spiral galaxy due to its staggering age of approximately 11 billion years. This revelation places it among the early inhabitants of the universe, providing a rare glimpse into the conditions that prevailed during the cosmic dawn.

The significance of A1689B11 lies not only in its age but also in its potential to unlock the secrets of galactic evolution. Ancient spiral galaxies challenge conventional wisdom about the timeline of spiral arm formation and the factors influencing their persistence over cosmic epochs.

As astronomers examine A1689B11, they seek clues about the conditions that allowed spiral galaxies to emerge in the early universe. The study of ancient spirals contributes to our understanding of the cosmic environment, the prevalence of specific galactic structures, and the role spiral galaxies played in shaping the cosmic landscape during the universe's infancy.

A1689B11 serves as a celestial time capsule, offering a unique perspective on the ancient epochs of galactic evolution. The exploration of this ancient spiral galaxy expands our knowledge of the diverse trajectories galaxies traverse across the vast cosmic expanse.

The Milky Way's Bar Structure

Amidst the myriad galaxies scattered across the cosmos, our very own Milky Way takes center stage, revealing a distinctive feature that sets it apart—the presence of a bar structure. This characteristic bar, composed of stars arranged in a straight bar shape, adds an intriguing layer to the narrative of our galactic home.

The Milky Way's bar structure distinguishes it as a barred spiral galaxy, a classification shared by approximately two-thirds of spiral galaxies. Unlike the arms that lead directly into the galactic center, the bar extends across the central region, connecting to the ends of the spiral arms. At the heart of this bar lies the galactic nucleus, a hub of stellar density.

The presence of a bar in the Milky Way contributes to its overall galactic dynamics. Bars play a role in redistributing mass within the galactic disk, influencing the motion of stars and gas. Understanding the impact of the bar on the Milky Way's structure provides valuable insights into the processes governing the evolution of barred spiral galaxies.

Observations of the Milky Way's bar structure reveal a complex interplay between stellar components and galactic morphology. The intricacies of this bar, coupled with the broader context of spiral arm dynamics, contribute to the ongoing exploration of our galactic neighborhood.

As we unravel the mysteries of the Milky Way's bar structure, we gain not only a deeper understanding of our galactic home but also insights into the broader classification of spiral galaxies. The Milky Way, with its distinctive features, invites us to contemplate the cosmic symphony playing out within its spiral arms and bar structure.

Classification of Spiral Galaxies

As we navigate the celestial expanse, the diverse array of spiral galaxies beckons us to explore the intricacies of their classification. In this chapter, we delve into the classification of spiral galaxies, unveiling the four main galaxy classes, peering into the unique realm of barred spiral galaxies, and acknowledging the challenges astronomers face in categorizing these captivating cosmic entities.

Four Main Galaxy Classes

The universe's vast tapestry reveals an astonishing diversity of galaxies, each characterized by unique features and structures. Astronomers have established a classification system to categorize galaxies based on their observable traits. Within this taxonomy, spiral galaxies represent a significant category, further subdivided into four main classes: spiral, barred spiral, elliptical, and irregular.

Spiral Galaxies: The Cosmic Twisters
Spiral galaxies, with their captivating spiral arms and central bulges, constitute one of the primary classes. These galaxies feature a central bulge surrounded by a rotating disk,

from which spiral arms emanate. The Milky Way serves as an exemplary specimen of a spiral galaxy, showcasing the intricate dance of stars within its rotating disk and spiral arms.

Barred Spiral Galaxies: Cosmic Bars in the Galactic Dance

Approximately two-thirds of all observed spiral galaxies boast an additional feature—a central bar structure. Barred spiral galaxies present a distinctive appearance, with a straight bar of stars connecting the ends of the spiral arms. This bar, often housing the galactic nucleus, introduces an additional layer of complexity to the galactic morphology.

Elliptical Galaxies: The Rounded Enigmas

Elliptical galaxies, in contrast to the structured spirals, exhibit a rounded, elliptical shape. These galaxies lack the prominent spiral arms and central bulge observed in spirals, consisting primarily of older, dimmer stars. Their elliptical form suggests a more settled state compared to the dynamic spirals.

Irregular Galaxies: Cosmic Chaos

Irregular galaxies defy the ordered patterns of spirals and ellipticals. These galaxies exhibit a chaotic arrangement of stars, gas, and dust, lacking the symmetrical structure observed in their counterparts. Irregular galaxies contribute a sense of unpredictability to the cosmic narrative.

This four-fold classification system provides a framework for understanding the diversity of galaxies across the cosmos. Within the realm of spirals, the presence or absence of a central bar contributes to further subdivisions, offering astronomers a nuanced language to describe the cosmic variety.

Barred Spiral Galaxies

Barred spiral galaxies, a subset of spiral galaxies, captivate astronomers with their unique structures and dynamic features. The defining characteristic of these galaxies is the presence of a central bar—a straight bar-shaped concentration of stars that connects the spiral arms. This bar structure introduces a level of complexity to the traditional spiral appearance.

Structure and Dynamics of Barred Spirals

The central bar in barred spiral galaxies often extends across a significant portion of the galactic disk. From the ends of this bar, the familiar spiral arms emerge, creating a visually distinctive pattern. The galactic nucleus, a region of heightened stellar density, is typically located at the center of the bar.

The gravitational influence of the central bar plays a crucial role in shaping the dynamics of barred spiral galaxies. It redistributes mass within the galactic disk, affecting the motion of stars and gas. The interaction between the bar, spiral arms, and the surrounding galactic environment contributes to the overall morphology and evolution of these cosmic entities.

Prominent Examples of Barred Spirals
Observations of the cosmos reveal a multitude of barred spiral galaxies, each with its unique characteristics. Notable examples include NGC 1300, a barred spiral galaxy in the constellation Eridanus, and NGC 1365, a barred spiral galaxy in the Fornax cluster. These galaxies showcase the elegance and complexity introduced by the presence of a central bar.

The Milky Way, our cosmic home, is also classified as a barred spiral galaxy. The recognition of a central bar in the Milky Way adds a layer of fascination to our understanding of the galactic neighborhood. Studying the dynamics of the Milky Way's bar structure provides insights into the broader behavior of barred spirals across the universe.

Galactic Evolution and Bars

The role of bars in the evolution of galaxies remains an active area of research. Galactic mergers, where two galaxies interact gravitationally and potentially form a larger barred spiral, represent a plausible mechanism for the creation and evolution of bars. Understanding the factors influencing the formation and persistence of bars contributes to our comprehension of galactic dynamics over cosmic timescales.

Challenges in Classifying Spiral Galaxies

While the classification of spiral galaxies offers a structured framework, astronomers face challenges in precisely categorizing these celestial entities. The appearance of spiral galaxies can vary considerably depending on their orientation relative to Earth, presenting challenges in accurately discerning their features.

Orientation Challenges

Classifying spiral galaxies is not always straightforward, as their appearance changes based on our vantage point. A galaxy viewed face-on, where the bulge and all spiral arms are clearly visible, presents a different visual profile than one seen edge-on, where only the outer edge of one side of the arms is observable. The variation in orientation complicates the classification process, leading to potential misinterpretations of a galaxy's true structure.

Diverse Morphologies

Spiral galaxies, even within the same class, exhibit diverse morphologies. The appearance of spiral arms, the size of the central bulge, and the presence or absence of a bar can vary significantly from one galaxy to another. The challenge lies in developing a classification system that accommodates this diversity while providing meaningful insights into galactic structure and evolution.

Hybrid Structures

Some galaxies defy easy categorization, displaying hybrid structures that blend characteristics of different classes. For example, a galaxy might exhibit both spiral arms and a central bar, blurring the lines between traditional classifications. These hybrids pose challenges in creating a clear taxonomy and highlight the need for a flexible classification system that accommodates the richness of observed galactic morphologies.

Despite these challenges, the ongoing advancements in observational technology and data analysis contribute to refining the classification of spiral galaxies. Innovative techniques, including detailed imaging and spectroscopy, allow astronomers to glean deeper insights into the structures and dynamics of these galaxies, paving the way for a more nuanced understanding of their cosmic diversity.

The Milky Way and Webb Telescope

In the cosmic expanse, our home galaxy, the Milky Way, beckons us to explore its intricate features and hidden mysteries. In this chapter, we embark on a cosmic journey, unraveling the characteristics of the Milky Way and delving into the transformative role of the James Webb Space Telescope in advancing our understanding of spiral galaxies, including our very own.

The Milky Way's Features

The Milky Way, a vast spiral galaxy, unfolds above us as a celestial river, weaving through the night sky. As we gaze upon its brilliance, the Milky Way reveals a tapestry of features that captivate astronomers and stargazers alike.

Spiral Structure and Galactic Components

At the heart of the Milky Way lies a complex spiral structure, characterized by a central bulge and rotating disk. The central bulge, composed of older, dimmer stars, conceals the potential presence of a supermassive black hole—a gravitational giant that influences the surrounding galactic environment.

The rotating disk, where the majority of stars reside, showcases the distinctive spiral arms that wind their way through the galactic expanse. These spiral arms, adorned with gas, dust, and young stars, contribute to the cosmic beauty of the Milky Way. The interplay between the central bulge, rotating disk, and spiral arms creates a dynamic environment that shapes the galaxy's evolution.

Beyond the disk, the Milky Way extends into sparsely populated halos—roughly spherical regions above and below the plane of the galactic disk. These halos, often hosting globular clusters and dark matter, add another layer to the multifaceted structure of our galactic home.

The Milky Way's Bar Structure

Adding to the complexity of the Milky Way is its classification as a barred spiral galaxy. Unlike traditional spirals, the Milky Way features a central bar—a straight bar-shaped concentration of stars that connects the ends of the spiral arms. This bar structure, housing the galactic nucleus, enhances the galactic dynamics and contributes to the overall morphology of our cosmic abode.

The recognition of the Milky Way's bar structure, which became more evident through observational advancements, has deepened our understanding of our galactic neighborhood. Studying the dynamics of the Milky Way's bar provides valuable insights into the behavior of barred spiral galaxies—a class that represents a significant portion of observed spirals.

Webb Telescope's Contribution to Spiral Galaxy Research

As humanity's quest to understand the cosmos continues, the James Webb Space Telescope (JWST) emerges as a transformative tool, poised to revolutionize our exploration of spiral galaxies, including the Milky Way. In this section, we explore the capabilities of the Webb Telescope and its potential contributions to unraveling the secrets of spiral galaxy research.

Webb Telescope Overview

The James Webb Space Telescope, often dubbed as the successor to the Hubble Space Telescope, represents a leap forward in observational capabilities. Positioned at the second Lagrange point (L2), approximately 1.5 million kilometers from Earth, the Webb Telescope offers a unique vantage point for exploring the universe.

Equipped with a suite of advanced instruments, the Webb Telescope operates primarily in the infrared part of the electromagnetic spectrum. This infrared focus allows astronomers to peer through cosmic dust, revealing hidden details of celestial objects and enabling observations of distant galaxies, stars, and planetary systems.

Peering Through Cosmic Dust

One of the challenges in studying spiral galaxies lies in the presence of cosmic dust that obscures visible light observations. The Webb Telescope's infrared capabilities provide a solution to this challenge, allowing astronomers to penetrate the dust clouds that often shroud the inner regions of spiral galaxies.

By observing in the infrared, the Webb Telescope unveils the obscured regions of galactic centers, where the interplay between stars, gas, and potential supermassive black holes influences galactic dynamics. This capability is particularly crucial for understanding the intricate structures within barred spiral galaxies, where the bar plays a significant role in shaping the galactic environment.

Studying Star Formation in Spiral Arms

Spiral arms are key regions of star formation within galaxies. The Webb Telescope's sensitivity to infrared radiation enables detailed studies of these stellar nurseries. By focusing on the infrared signatures associated with young, hot stars and the surrounding dust clouds, astronomers can gain insights into the mechanisms driving star formation in spiral arms.

Understanding the processes behind star formation in spiral arms contributes to our broader comprehension of galactic evolution. The Webb Telescope's ability to peer into these regions, unaffected by the dust that hinders visible light observations, opens a new frontier in unraveling the mysteries of stellar birth within spiral galaxies.

Probing Galactic Dynamics and Evolution
The study of galactic dynamics, including the role of bars in spiral galaxies, benefits significantly from the Webb Telescope's capabilities. Infrared observations allow astronomers to trace the motion of stars and gas within galactic disks, providing a deeper understanding of the forces at play in shaping the structure of spiral galaxies.

Additionally, the Webb Telescope's capacity for detailed spectroscopy facilitates the analysis of galactic chemical compositions.

By studying the elemental abundances within spiral galaxies, astronomers can piece together the galactic history and trace the evolutionary pathways that lead to the diverse morphologies observed in the universe.

Exploring the Milky Way's Galactic Structure

While the Webb Telescope's primary focus is on distant galaxies, its capabilities extend to the study of our own Milky Way. Infrared observations allow astronomers to peer through the dense regions of the galactic plane, revealing obscured features such as the Milky Way's central bar and providing a clearer view of the galactic nucleus.

Mysteries of Spiral Galaxies

In the cosmic panorama, spiral galaxies stand as enigmatic cosmic entities, veiled in mysteries that beckon astronomers to unravel their secrets. In this chapter, we delve into the perplexing mysteries surrounding spiral galaxies, exploring the puzzle of spiral arm stability and the ongoing research and discoveries that promise to illuminate the intricate tapestry of these celestial wonders.

Spiral Arm Stability Puzzle

The captivating spiral arms of galaxies, though visually stunning, present astronomers with a profound puzzle — the mystery of spiral arm stability. Unlike static structures, these arms exhibit dynamic, winding patterns that persist over cosmic timescales. The question of how spiral galaxies maintain the stability of their intricate arms remains a central enigma in modern astrophysics.

Dynamism in Spiral Arm Formation

Spiral arms, characterized by concentrations of stars, gas, and dust, create the iconic spiral patterns that define galaxies. The challenge lies in understanding the dynamic processes that govern the formation, persistence, and evolution of these arms.

One prevailing theory proposes the existence of density waves traversing the galactic disk as the orchestrators of spiral arm formation. These density waves, akin to ripples in a celestial pond, travel through the outer disk of the galaxy, compressing and triggering the formation of spiral patterns. However, the inherent dynamism of these waves raises questions about how spiral arms maintain their structure over the vast timescales of galactic evolution.

The Role of Galactic Rotation

A key factor in the stability of spiral arms is the rotation of the galaxy itself. As stars orbit the galactic center, they move through the spiral arms, influenced by the gravitational forces at play. The combination of the rotation of the galaxy and the density waves creates a dynamic environment, with stars moving in and out of the arms.

Understanding the interplay between galactic rotation and density waves is essential for deciphering the stability puzzle. The challenge lies in reconciling the transient nature of individual stars within the arms with the persistence of the overall spiral structure. Astronomers grapple with questions about how the rotating disk and density waves synchronize to maintain the intricate arms that define spiral galaxies.

Cosmic Recycling and Star Formation

Another layer of complexity in the stability puzzle involves the continuous process of star formation within spiral arms. As new stars emerge, they contribute to the luminosity and dynamics of the arms. Yet, these stars follow their own life cycles, evolving from hot, young stars to older, dimmer ones.

The recycling of stellar material within spiral arms adds a temporal dimension to the stability puzzle. As stars age and transition through their life cycles, their gravitational interactions with the surrounding environment may influence the overall structure of the arms. The intricate dance of star formation, evolution, and gravitational forces complicates the task of unraveling the mysteries of spiral arm stability.

Ongoing Research and Discoveries

In the quest to decipher the mysteries of spiral galaxies, ongoing research and technological advancements promise to unveil new insights, pushing the boundaries of our understanding of these cosmic wonders.

Webb Telescope's Gaze into Spiral Galaxies
The James Webb Space Telescope emerges as a game-changer in unraveling the mysteries of spiral galaxies. Equipped with

advanced infrared capabilities, the Webb Telescope offers a unique vantage point for peering through the cosmic dust that often obscures visible light observations. This capability is crucial for studying the central regions of spiral galaxies, where the dynamics of spiral arms and potential supermassive black holes unfold.

By focusing on infrared observations, the Webb Telescope enables astronomers to delve into the heart of spiral galaxies, exploring the regions where density waves, star formation, and galactic rotation converge. The telescope's ability to provide detailed spectroscopy and high-resolution imaging promises to enhance our understanding of the factors influencing spiral arm stability.

Simulations and Computational Models

Advancements in computational modeling and simulations play a pivotal role in tackling the complexities of spiral arm stability. Astrophysicists harness the power of supercomputers to simulate the gravitational interactions, gas dynamics, and stellar evolution within galactic disks.

These simulations offer a virtual laboratory where researchers can experiment with various parameters to understand the conditions that lead to the emergence and persistence of spiral arms. By incorporating realistic models of star formation, feedback mechanisms, and galactic environments, scientists aim to bridge the gap between theoretical predictions and observed galactic structures.

Galactic Surveys and Big Data

The era of big data in astronomy has ushered in a wealth of observational information through extensive galactic surveys. Projects like the Sloan Digital Sky Survey and the upcoming Large Synoptic Survey Telescope (LSST) are poised to revolutionize our understanding of galactic structures, including spiral arms.

These surveys, equipped with advanced imaging capabilities, capture unprecedented amounts of data, providing a comprehensive view of the diverse morphologies within spiral galaxies. The analysis of these datasets allows astronomers to identify patterns, correlations, and anomalies, offering valuable clues about the underlying mechanisms that govern spiral arm stability.

Interdisciplinary Approaches

The study of spiral galaxies benefits from interdisciplinary collaboration, drawing insights from fields such as fluid dynamics, plasma physics, and complex systems theory. Researchers explore analogs in laboratory experiments, seeking to replicate the conditions that may influence the stability of spiral arms.

By combining theoretical astrophysics with experimental approaches, scientists aim to refine our understanding of the physical processes governing galactic dynamics. The interdisciplinary synergy brings fresh perspectives to the stability puzzle, fostering innovative solutions and novel avenues for exploration.

Machine Learning and Pattern Recognition
The rise of machine learning and artificial intelligence introduces powerful tools for analyzing the intricate patterns within spiral galaxies. Machine learning algorithms can sift through vast datasets, identifying subtle features, correlations, and anomalies that may escape human observation.

These algorithms contribute to automating the process of classifying and characterizing galactic structures. As machine learning techniques evolve, they hold the potential to uncover hidden patterns within spiral arms, shedding light on the underlying mechanisms that contribute to their stability.

Webb Telescope's Observations

As humanity's cosmic eye, the James Webb Space Telescope (JWST) transcends the boundaries of its predecessors, offering unprecedented insights into the hidden realms of the universe. In this chapter, we explore the transformative observations of the Webb Telescope, focusing on its prowess in mid-infrared and near-infrared imaging and the profound insights gleaned from its scrutiny of specific galaxies.

Mid-infrared and Near-infrared Imaging

The James Webb Space Telescope's observational capabilities extend into the realms of mid-infrared and near-infrared imaging, opening new vistas for astronomers eager to unravel the mysteries of the cosmos. These specific regions of the electromagnetic spectrum provide a unique lens through which to peer into the intricate details of celestial objects, offering advantages that complement and surpass the capabilities of its predecessor, the Hubble Space Telescope.

Infrared Advantage: Peering Through Cosmic Dust

The mid-infrared and near-infrared regions of the spectrum hold a distinct advantage over visible light observations, particularly when it comes to penetrating the cosmic veils of dust that shroud various celestial structures. Dust, while contributing to the birth of stars, often obscures our view of galaxies, star-forming regions, and other cosmic phenomena in visible light. The Webb Telescope's ability to operate in the infrared allows astronomers to see through this cosmic haze, revealing hidden details that were once elusive.

Infrared observations are particularly crucial for studying the central regions of spiral galaxies, where dense concentrations of stars, gas, and potential supermassive black holes reside. By harnessing the power of infrared light, the Webb Telescope provides a clearer view of these galactic cores, allowing astronomers to explore the dynamics of spiral arms, the interplay of stars, and the presence of central bars in unprecedented detail.

Spectroscopic Capabilities: Unraveling Galactic Chemistry

Beyond imaging, the Webb Telescope's advanced spectroscopic capabilities further enhance our understanding of the chemical compositions within galaxies. Spectroscopy involves analyzing the light emitted or absorbed by celestial objects, providing a wealth of information about the elements present, their abundances, and the physical conditions of the observed regions.

In the context of spiral galaxies, spectroscopy allows astronomers to unravel the galactic chemistry embedded within the luminous tapestry of stars and gas. By examining the infrared signatures of different chemical elements, researchers can deduce the elemental compositions of stars, gas clouds, and various structures within spiral arms. This chemical census contributes to our comprehension of the evolutionary processes shaping spiral galaxies over cosmic timescales.

Insights from Webb's Observations of Specific Galaxies

The Webb Telescope's vigilant gaze extends to specific galaxies, unraveling their unique stories and providing profound insights into the dynamic interplay of celestial forces. Through targeted observations, astronomers have uncovered hidden details, refined existing models, and deepened our appreciation for the diverse morphologies exhibited by spiral galaxies.

NGC 7496: A Barred Spiral Unveiled

NGC 7496, a barred spiral galaxy located in the constellation Grus, emerged as a celestial canvas for the Webb Telescope's scrutiny. This galaxy, characterized by its central bar structure, presented an opportunity to delve into the intricacies of barred spirals and refine our understanding of their dynamics.

Webb's mid-infrared and near-infrared imaging capabilities unveiled the central bar in NGC 7496 with unprecedented clarity. The telescope's high-resolution observations allowed astronomers to trace the bar's structure and discern the intricate patterns within its stellar congregation. Insights gained from NGC 7496 contribute not only to our understanding of individual barred spirals but also to the broader classification and evolution of spiral galaxies.

NGC 1433: A Barred Spiral in Detail

NGC 1433, another captivating barred spiral galaxy located in the constellation Horologium, became the focus of Webb Telescope's meticulous observations. This galaxy, with its central bar and spiral arms, provided an opportunity to explore the intricate dynamics of a barred spiral on a finer scale.

Webb's imaging capabilities captured the details of NGC 1433's central region, unraveling the subtle features within the bar and the surrounding stellar environment. The telescope's ability to peer through cosmic dust enhanced the clarity of the observations, allowing astronomers to study the bar's influence on star formation and gas dynamics. Webb's observations of NGC 1433 contribute valuable data to the ongoing quest to understand the role of bars in shaping the morphology and evolution of spiral galaxies.

M74: Grand Design Spiral Unveiled

M74, a grand design spiral galaxy located in the constellation Pisces, stands as an exemplary specimen of spiral elegance. Webb Telescope's gaze toward M74 offered a unique opportunity to explore the characteristics of grand design spirals — galaxies with prominent and regular spiral arms as compared to their patchier counterparts.

The mid-infrared and near-infrared imaging capabilities of the Webb Telescope allowed astronomers to study the grand design of M74 in exquisite detail. The telescope's observations highlighted the symmetrical arms and the central region, shedding light on the factors influencing the formation and persistence of such striking spiral patterns. Webb's scrutiny of M74 enriches our understanding of the diverse morphologies within the category of spiral galaxies, showcasing the beauty and complexity inherent in grand design spirals.

Stephan's Quintet: A Cosmic Ensemble in Infrared Harmony

Webb Telescope's gaze extended beyond individual galaxies to capture the intricate dance of multiple galaxies in Stephan's Quintet. This cosmic ensemble, located in the constellation Pegasus, consists of five galaxies engaged in complex gravitational interactions. Webb's mid-infrared and near-infrared imaging not only revealed the individual structures of the galaxies but also highlighted the interplay of hot dust and newly-formed stars.

In this observation, NIRCam highlighted the bright and newly-formed stars, while MIRI unveiled the hot dust and broader galactic structure. The combined imaging offered a multi-dimensional view of the interactions within Stephan's Quintet, providing astronomers with valuable data to study galactic mergers, star formation triggered by gravitational interactions, and the overall dynamics of galactic ensembles.

Insights into Spiral Galaxy Evolution

The targeted observations of specific galaxies by the Webb Telescope contribute to our broader understanding of spiral galaxy evolution. By examining the detailed structures, dynamics, and chemical compositions within these galaxies, astronomers piece together the cosmic narratives that unfold over cosmic timescales.

Webb's observations serve as a testament to the transformative capabilities of advanced space telescopes, offering a front-row seat to the cosmic drama playing out within spiral galaxies. As the telescope continues to peer into the depths of the universe, its contributions to unraveling the mysteries of specific galaxies and refining our understanding of their celestial intricacies mark a new chapter in observational astronomy. The insights gained from Webb's observations transcend individual galaxies, shaping our comprehension of the broader cosmic tapestry in which spiral galaxies are integral threads.

Conclusion

As we draw the cosmic curtains on our exploration of spiral galaxies, this concluding chapter reflects on the key insights gleaned from our journey and peers into the celestial horizon to contemplate future directions in spiral galaxy research.

Recap of Key Insights

Our celestial voyage through the realms of spiral galaxies has unveiled a tapestry of insights, each thread contributing to our understanding of these cosmic wonders.

Diversity of Galactic Structures

Spiral galaxies, with their mesmerizing spiral arms, central bulges, and diverse morphologies, stand as a testament to the cosmic diversity that populates our universe. From grand design spirals like M74 to barred spirals such as NGC 7496, each galaxy tells a unique story shaped by gravitational interactions, stellar dynamics, and the interplay of cosmic forces.

Central Bulge and Rotating Disk Dynamics
The interplay between the central bulge and rotating disk within spiral galaxies forms the core of their dynamics. The presence of a central bulge, often harboring a supermassive black hole, influences the motion of stars within the galactic disk. This

dynamic dance shapes the overall morphology of spiral galaxies, with the intricate spiral arms emerging as a result of density waves traveling through the galactic expanse.

Formation and Evolutionary Pathways

The puzzle of spiral arm formation continues to intrigue astronomers, with density waves and galactic encounters emerging as leading contenders in explaining the emergence of these cosmic features. The evolutionary trajectory from spiral to elliptical galaxies remains a tantalizing mystery, with ongoing research aiming to unravel the factors that drive galactic metamorphosis.

Notable Galaxies and Classification Challenges

NGC 6872's expansive spiral arms, A1689B11's ancient cosmic tale, and the Milky Way's bar structure serve as celestial landmarks within the vast landscape of spiral galaxies. The classification of these cosmic entities, whether as grand design spirals, barred spirals, or hybrids, poses challenges as astronomers navigate the complexities of varying orientations and morphologies.

Webb Telescope's Transformative Role

The James Webb Space Telescope emerges as a celestial sentinel, peering into the heart of spiral galaxies with its mid-infrared and near-infrared imaging capabilities. From unraveling the mysteries of specific galaxies like NGC 7496 and NGC 1433 to capturing the cosmic ensemble in Stephan's Quintet, Webb's observations provide a nuanced view of the structures, dynamics, and evolutionary pathways within spiral galaxies.

Mysteries and Ongoing Research

The dynamism of spiral arm stability, the cosmic recycling of star formation, and the underlying processes influencing galactic evolution persist as mysteries awaiting further exploration. Ongoing research, propelled by advanced simulations, interdisciplinary approaches, and the ever-expanding realm of big data, promises to deepen our understanding of these cosmic enigmas.

Future Directions in Spiral Galaxy Research

As we gaze toward the future, the cosmic odyssey of spiral galaxy research opens new frontiers and beckons astronomers to push the boundaries of knowledge.

Advanced Observational Technologies

The evolution of observational technologies continues to play a pivotal role in advancing our understanding of spiral galaxies. Future telescopes, equipped with even more advanced instruments and capabilities, will allow astronomers to probe deeper into the cosmic abyss. The development of space-based and ground-based observatories, spanning a range of wavelengths, promises to unveil previously unseen details within spiral galaxies.

Multi-Wavelength Surveys

The era of multi-wavelength surveys, driven by projects like the Large Synoptic Survey Telescope (LSST), heralds a new era in galactic exploration. These surveys, capturing extensive datasets across the electromagnetic spectrum, will provide a comprehensive view of the diverse morphologies, chemical compositions, and dynamics within spiral galaxies. By combining data from different wavelengths, astronomers can construct a holistic understanding of the cosmic processes shaping these celestial entities.

Interdisciplinary Collaboration

The mysteries of spiral galaxies transcend traditional astronomical boundaries, inviting collaboration with experts from diverse scientific disciplines. Fluid dynamics, plasma physics, and complex systems theory converge with astrophysics to unravel the intricate dynamics of spiral arm stability. Interdisciplinary approaches, combining theoretical insights with laboratory experiments, hold the promise of innovative solutions to longstanding puzzles.

Advancements in Computational Modeling
The computational frontier continues to expand, offering astrophysicists increasingly sophisticated tools to simulate the

complexities of spiral galaxy dynamics. High-performance computing allows researchers to model gravitational interactions, gas dynamics, and star formation processes within galactic disks. Advanced simulations, incorporating realistic scenarios and feedback mechanisms, pave the way for a more nuanced understanding of the factors influencing galactic evolution.

Machine Learning and Artificial Intelligence The rise of machine learning and artificial intelligence introduces a transformative dimension to spiral galaxy research. Machine learning algorithms, capable of analyzing vast datasets and identifying subtle patterns, accelerate the pace of discovery. These algorithms aid in the automated classification of galaxies, the recognition of complex structures, and the extraction of meaningful insights from the wealth of observational data.

Long-Term Observational Campaigns
Spiral galaxies, with their evolving structures and dynamic processes, benefit from long-term observational campaigns. Continuous monitoring of specific galaxies over extended periods allows astronomers to capture rare events, trace the evolution of galactic structures, and unravel the temporal dynamics within these celestial entities. Long-term campaigns provide a dynamic window into the ongoing saga of spiral galaxies.

Exploration of Galactic Environments

The study of spiral galaxies extends beyond individual entities to the exploration of their cosmic neighborhoods. Investigating the influence of galactic environments, interactions between galaxies, and the role of dark matter in shaping spiral structures opens avenues for a comprehensive understanding of the broader cosmic context. Observations of galaxy clusters and cosmic filaments contribute to unraveling the intricate dance of gravitational forces shaping the cosmic tapestry.

In conclusion, the journey through the cosmic realms of spiral galaxies is an ongoing odyssey of discovery and contemplation. From the intricate structures within individual galaxies to the cosmic ballet of galactic ensembles, the mysteries that unfold offer a glimpse into the profound complexities of the universe. As astronomers continue to explore, innovate, and collaborate

www.ingramcontent.com/pod-product-compliance
Lightning Source LLC
Chambersburg PA
CBHW071057290526
45795CB00004B/1536